Internet Homesteading
How To Make Your Own Internet Connection

I0478422

Table of content

Introduction

You lead an adventurous life. You don't have time for all the hustle and bustle of the trending world, you have places to go and sights to see. As you browse through the internet pages, you only fuel the fire for wanting to go more places, do more things, and meet more people.

But, you find when you venture out into the wild, you quickly lose the one thing you rely on more than you realize – the internet. You never think about how much you rely on it, until you suddenly find yourself without it. It is then that you realize that you need to have access – but as you look around, you wonder how you are going to make that happen.

There was a time, of course, when you didn't have a reliable cell phone, and you would have to walk around the room (or the campsite) in order to find a connection to make calls. As you venture out into the world, you realize that you can quickly lose your internet connection – cutting you off from valuable information you may need.

Let's face it, in this modern world we live in, the internet truly is the source for the greatest variety of information. Whether you are looking up a recipe, you are looking up how to create a project, or you are looking up something that is health related, you need to have a connection to the online community regardless of where your adventures take you.

Sure, when you are living in the heart of a city you are able to find this kind of information easily, but what about when you are in the heart of the forest, and you realize that you need help finding which direction you are heading? Or when you are out camping and you find that you need to come up with an alternative to the shelter you have brought with you?

There are countless situations you may find yourself in, and you will realize that you need to get online in a moment's notice. It doesn't have to be an emergency, but it could be. There doesn't have to be a dangerous reason you are off the grid, but there could be.

There doesn't have to be any reason at all why you need to get online, and get online now – but regardless of the life you lead, knowing how to get connected in a heartbeat is going to give you a peace of mind that you have never before felt.

So are you ready to ensure that you will always have internet access? Good, let's get started.

Chapter 1 – What's the Big Deal?

In this modern world that we live in, it's easy to forget how reliant we are on technology. We expect things to work when we need them to, how we need them to, and whenever we want them to. When they don't, tech support is just a phone call away, and in no time at all you are back in the virtual world, doing what you have always done.

After all, with the internet's features, your day is far easier than it used to be. With each passing day, in fact, more information is added to the cloud, and more information is made directly available to you when you need it. It doesn't matter where you are on the planet, if there is something you want to know, it's made available to you within seconds of you wanting to know it.

In fact, the term "Google it" has become so commonplace that no one gives it a second thought when there is a question that no one knows the answer to. Then, with a few taps on the touch screen, the information is there and ready to be used.

Although it might seem to be a contradiction that someone who wishes to be off the grid would use the internet, it's important that you do have internet access (or at least know how to get that access) when you are in need of doing so. Even those who are only online once every few days still need the internet's service, though they aren't using it often or for the same reasons that many other people are using it.

Our modern world tends to associate the internet with social media and the need to always be streaming music and videos. While these are great bonuses to having the internet, the fact is the internet can be used for many other useful purposes that come in handy in a variety of situations.

The internet erases the lines of cultures and peoples, meaning no matter where you are in the world, if you have internet access, you are able to keep in contact and stay up to date on what else is going on in the world.

Let's face it. Though many people claim they are able to get offline at any time and be perfectly happy in life, the internet has become such a crucial part of our society, it's almost necessary to stay connected in at least a minimal form. Even if you don't use the internet as a way to keep in touch with friends, family, or even to make connections with the world around you, you do need to keep up with the current events in the news.

It doesn't matter if you are going off the grid simply because you want to be less involved with the modern day society, or if you are forced off the grid due to some unforeseen event or natural disaster, when you are unable to access the internet, you cut yourself off from what is going on in the world.

In another sense, keeping internet access is so important for the individual because the world is constantly changing. With new technology being introduced on a daily basis, it is crucial to know what is going on in order to keep up with the times.

By going off the grid completely for even a month things could change drastically for the individual, making re-entry into the modern world an incredibly difficult process. Though one may only wish to be part of the modern world for a short period of time, it is crucial that each individual in societies across the world know how to use digital devices and understand how to use digital resources.

So, with internet access being such a crucial thing in the modern world that we live in, what can you do to ensure you are always going to have the access you need to function?

Though we live in a world in which so many different devices can access the world wide web, there are still variables that you need to take into account when you are learning how to maintain a constant access to the internet. Think about it, if you were to rely solely on your cell phone for internet access, what are you going to do if the battery is dead and you have no way to charge it?

Or what will you do when you are out of cell phone service areas, and have no way of returning to a service area at the moment? Suppose you rely on public computers for your internet access – what are you going to do if natural disaster strikes and you have no way to get to the public place to use the computer?

Suppose you get lost and have no means by which to get online, simply because you didn't choose to have a device that is capable of the access?

While some situations might seem a little far-fetched, it is always good to be prepared. You never know when you are going to find yourself in a situation in which you need to get online, and you never know what means you will have to do so when the need arises.

But, there is good news. With the right mindset, the right tools, and the right set of skills, you can enable yourself to get online from virtually anywhere. By using the right means, you can constantly be connected with the world around you, even when you are entirely off the grid. With this capability, you give yourself the opportunity to stay in contact with those you wish to stay in contact with, you are able to keep current not only with the latest world and news events, but also with the latest digital technology.

And, you are able to have access to countless resources of information, allowing yourself to always be able to have the answer when you need it. With all these crucial benefits of internet access, what are you waiting for? In the chapters to come, we are going to look at how you can gain this internet access any time you need to, in any way you need to.

Yet another way to keep the world at your fingertips.

Chapter 2 – Preparation is Key

When it comes to off the grid internet access, having the right tools on hand is going to make all the difference in how well you are able to connect to the online world. Let's start by looking at the different options you have to get online:

- **Cellular devices**

- **Tablets**

- **Laptop computers**

- **Desktop computers**

- **Smart devices**

Odds are, you have one, several, or even all these things in your home already. Perhaps you use them all frequently, or perhaps you have your favorites and rarely us the others. However, if you pay attention to this list of internet capable devices, you will notice that they have something in common – they all require electricity to work.

Sure, most of these devices do run off of battery power, but you are going to have to charge that battery at some point if you want the device to work. With this in mind, you are going to have to keep the right tools on hand in your home at all times.

These include:

- Charging cables

- Batteries

- Battery powered charging stations (with batteries that will work for these stations)

- Backup devices (in the event something has happened to your current device)

- A router

- A modem

- An ethernet cable

- Ham radio (available in many technical supply stores, as well as Amazon)

A modem and a router are the two tools that are needed to get your devices connected to the internet. A router has a specific port for the ethernet cable, and while it's not entirely necessary for you to have an ethernet cable to get online, you will get a much stronger connection from the particular device you plug into the router directly through this cable.

When you are in an emergency situation (or even if you are living in a remote area in which internet access is limited) you will want to concentrate as much power as possible to the device you are using to get online. With an ethernet cable, the device that is plugged into the router is given top priority for the internet, and the connection will be far stronger.

In this modern world we live in, cell phones are incredibly easy to come by. Since there are many options that don't require any form of

payment plan, you have the option to purchase an extra for an emergency box.

By purchasing an extra cellular device (make sure it has internet capabilities before you buy) you have a backup in the event something goes wrong with your current device. Though you may not plan on anything happening to your current device, it is wise to have something else on hand just in case.

When you are in emergency situations especially, you never know what could happen, and your phone could easily become lost or broken in the chaos. By having a backup, you guarantee that you are going to have something that can get online.

Ham radios are easy to find, as well as easy to use.

You can purchase one for around $100 from Amazon, or order one from a different supplier if you would rather. Along with the radio you will receive a guide on how to effectively use it, so in no time at all you will be able to use your radio effectively in any situation.

Ham radios are also often referred to as amateur radios due to the fact they are used for non-commercial purposes. The ham radio has been around longer than the internet, and it is still an incredibly reliable way to get online when you are unable to use conventional methods for any reason.

These radios are versatile, meaning you can use them for images, texts, voice communication, and data research though they are really just a little radio. They are managed by the International Amateur Radio Union which is officially organized in three regions, although there are working members for this union across the globe, making the ham radio a feasible option for maintaining both communication and open information with the rest of the world.

With a ham radio, you can rest assured you are going to know what is going on in the world at all times – regardless of what is going on in the world around you. This makes them a better option than computers in some ways, as you are going to be able to use them even during an emergency or natural disaster.

Though few people realize it, the ham radio is considered to be the most effective form of communication during an emergency.

Chapter 3 – Standard Procedures for Getting Online

Every true survivalist (or off-the-grid enthusiast) knows that in order to be successful in the situation you are in, you must gather the right tools for the situation. While you may not wish to get everything that is on the list to keep in your emergency (or functional) box, you should make sure that you have some sort of computer and a ham radio.

In addition to having the right kind of tools and devices for internet access, it's important to also important to understand what your options are when it comes to internet access.

In our modern world of wifi and high speed connections, we get so used to being able to get online from any room in the house, and enjoy any form of entertainment that we want. If you want to stream music you can, while in the other room someone is watching a movie and someone else is flipping through social media.

However, high speed, wifi internet is not the only internet that is available, and you may have more options than you realize.

We all remember the days of dial up internet. The days when you had to sit in front of the screen while the internet slowly loaded. Though this is not an ideal form of internet to choose if you are going to watch movies or videos, it is a perfectly feasible option for someone who is going to live off the grid as much as possible.

Dial up internet, unlike high speed internet, is connected to the landline (hence the reason phone calls wouldn't come through on the landline when you are online.) Though it's slower than high speed internet, it is more reliable in the sense that dial up internet is run by the phone companies – so if electricity has been knocked out due to a storm or some other natural disaster, but you still have landline service, you will be able to access the internet.

Keep in mind that while you are still going to have to pay for internet access through a landline, it is far less expensive than high speed internet. This means that it is still an option for those who are using it as a backup in case of emergency, as well as a good option for those who are trying to live off the grid.

As I mentioned, when it comes to the internet world, things are always growing and changing – and it is up to you to keep up with them.

The internet world isn't waiting for anyone, and as I already mentioned, if you are completely offline for any length of time, you run the risk of being left in the dark. Though we know we have devices that are capable of being online, as well as radios that can be used separate of (and at times, better than) the internet, and have compared the modern high-speed with the dial up that is still an option, that's not where it ends.

Wireless companies and internet service providers have joined forces, and though they are working to make what we have faster, more reliable, and better than ever, that is not the only thing they are working to accomplish. In fact, they are currently working on projects which will provide the entire nation with wifi services, regardless of where a router is.

Although there is a lot of concern with security in this idea, the internet continues to advance, and with it, the security will advance as well. This is going to make it possible to connect to the internet at any time, as long as you are within range of a person who has wifi hotspot.

These aren't the only people who are working to improve internet access across the globe, however, as governments are also working together with the United States on a project known as Space X.

Space X is a private space flight company, and they are working to initiate the launch of a variety of low flying satellites, which will, in turn, make it possible for the entire world to have easy internet access, regardless of where they are.

This is going to make it easier for you to have a connection to the internet, regardless of the situation you are in (provided you have the correct tools to make that happen.) And it will also make it easier for those who wish to live off the grid to have constant access to the online world. The purpose of this internet connection is to give people who are in remote locations access to connecting with each other, as well as enabling them to keep up with the current world events.

Though the concept of many internet satellites orbiting the Earth is an idea that is still working to come to fruition, this isn't the first time the concept of satellite internet has been on the market.

Satellite internet has been around for years now, and though it is still a relatively new thing in the internet world, it does provide constant access to those who have the means to reach it.

One of the biggest benefits of satellite internet is that the satellites aren't affected by anything that happens on the surface of the planet, meaning regardless of the natural disaster or emergency situation you may face, you are still going to get through to the internet.

Find what works for you, then put it to the test. You can always experiment and find what you like the best before you commit to any one way of doing things. Just make sure that you know how to access and navigate the internet through the devices you have in the event that something does go wrong in the future.

Chapter 4 – Procedures for Emergency Situations

When it comes to the word *emergency* many different things come to mind. Though we often think of an emergency as something big such as a natural disaster or a fire or something of that sort, there are other kinds of emergencies you may face when you are trying to live off grid but still maintain internet access.

The biggest emergency you are going to face is the potential for internet to be out for you personally. Perhaps there was a power surge, perhaps there was an issue in the region – whatever the reason, you aren't able to get online.

It's not the time to panic, however, as there are backup things you can do that will allow you to have internet access for the short term.

Go to a public setting where they either offer computers for the public to use – or head to a coffee shop where you can find the wifi for your device.

Are you concerned you are going to miss that business meeting? Don't panic – instead, go to the library or other place where the internet is offered to the public and do what you need to do there. Though this may not be a reliable way to take care of your internet needs all the time, it does offer you a second option when you have to get online as soon as possible.

Coffee shops and other local places like them have free wifi that is offered to the public, so all you need to do is bring in your computer or other device, and you are ready to get online when you need – just keep in mind that shared wifi connections are more susceptible to being hacked, and don't share private information (or access private accounts) when you are on these kinds of connections.

If you have no choice, you may need to do so anyway, but keep in mind that your security isn't as high when you are on public wifi.

Learn to be flexible – and an opportunist.

For someone who is living off the grid, it is no secret that you need to take your opportunities when they arise. You never know when you are going to get the chance again, so you better take care of what you need to take care of when you have the option in front of you.

With that in mind, use the internet when you have the chance, and don't worry about it when you don't have access. When you do have the opportunity to get online, check what is most important (and handle them) first, then move onto the more frivolous side of things. Just because you are online doesn't mean that you have to handle everything at once, and once you learn how to be opportunistic with your internet access, you will find that you don't miss it as much when you don't have access to it.

And always, have a backup plan.

Like I already mentioned, backup devices are going to work wonders to keep you on the internet. But don't stop there. Form a plan for what you will do if you are ever in need of getting on the internet and you don't have immediate access. Perhaps you have a plan worked out with the neighbor to get online at his house.

Perhaps you can go to a friend's place. It doesn't ultimately matter what your plan is, just make sure you have one.

Plan and prepare, and you are going to have the access you want, when you want.

Conclusion

There you have it, everything you need to know to get online – regardless of where your life takes you. Though you don't give much thought to the conveniences of your life now, a true adventurer knows that in order to be successful in life, they are going to have to be prepared for anything.

This book is designed to give you the confidence you need to access the internet no matter where you find yourself. It doesn't matter if you are in the middle of the wilderness or find yourself in a city that has been struck with natural disaster, you will always know how to get on the web and keep up with what is going on.

You will always have access to information you need to keep you and your loved ones safe, as well as information that will make your off-the-grid experience even better.

I know you will develop the skills you need to make this happen, and with these simple methods, you aren't ever going to have to worry that you won't be able to access the internet. All you need is the right equipment and the right mindset, and you are going to find that getting online is easier than you ever thought possible.

Enter the world with a newfound confidence, and you are going to discover that no matter where your adventures take you, you will have the power of information at your fingertips, every single time.

Now get out there and pursue your adventures with a passion, and discover how you can stay connected to the world, when you are hidden from the world's view.

Good luck and stay wild.

FREE Bonus Reminder

If you have not grabbed it yet, please go ahead and download your special bonus report *"DIY Projects. 13 Useful & Easy To Make DIY Projects To Save Money & Improve Your Home!"*

Simply Click the Button Below

OR **Go to This Page**

http://diyhomecraft.com/free

BONUS #2: More Free & Discounted Books or Products

Do you want to receive more Free/Discounted Books or Products?

We have a mailing list where we send out our new Books or Products when they go free or with a discount on Amazon. Click on the link below to sign up for Free & Discount Book & Product Promotions.

=> Sign Up for Free & Discount Book & Product Promotions <=

OR Go to this URL

www.ingramcontent.com/pod-product-compliance
Lightning Source LLC
Chambersburg PA
CBHW061238180526
45170CB00003B/1352